ACADEMIC PRESS, INC.
111 Fifth Avenue, New York, New York 10003

United Kingdom Edition published by
ACADEMIC PRESS, INC. (LONDON) LTD.
Berkeley Square House, London W1X 6BA

PRINTED IN THE UNITED STATES OF AMERICA

CONTENTS

The authors are indebted to Mr. Charles McClure, Mr. Stephen Young, and Mrs. Yuen-Han Kan for their assistance in preparing the solutions to these problems.

Solution to Problem 1.2

First we note that for $v = \text{const}$ we must have $c_p = c_v = c$. Indeed, from

$$dh = Tds + vdP$$

we find that

$$\left(\frac{\partial h}{\partial P}\right)_T = T\left(\frac{\partial s}{\partial P}\right)_T + v = -T\left(\frac{\partial v}{\partial T}\right)_P + v = v = \text{const} ,$$

because

$$\left(\frac{\partial s}{\partial P}\right)_T = \frac{\partial(s,T)}{\partial(P,T)} = \frac{\partial(v,P)}{\partial(P,T)} = -\left(\frac{\partial v}{\partial T}\right)_P .$$

By integration

$$h - h_o = v(P - P_o) + \phi(T)$$

so that

$$u - u_o = h - h_o - (P - P_o)v = \phi(T)$$

and

$$c_p = \left(\frac{\partial h}{\partial T}\right)_P = \frac{d\phi}{dT} = \left(\frac{\partial u}{\partial T}\right)_v = c_v = c \quad \text{(say)} .$$

Hence

$$h - h_o = v(P - P_o) + c(T - T_o) . \qquad \text{(a)}$$

Further

$$ds = \frac{du + Pdv}{T} = \frac{dh - vdP}{T} = \frac{d\phi}{T} = \frac{cdT}{T}$$

and

$$s - s_o = c\ell n \frac{T}{T_o} . \qquad \text{(b)}$$

The elimination of temperature from Equations (a) and (b) yields

$$h(s,P) = v(P - P_o) + cT_o(e^{s/c} - 1)$$

with $h_o = s_o = 0$ at $P = P_o$.

Solution to Problem 1.3

The solution to Problem 1.1 contains expressions for $u(T,x)$ and $s(T,x)$. Hence, the fundamental equation $f = f(T,x)$ is found by direct substitution into the definition $f = u - Ts$.

(a) $f(T,v) = c_v(T - T_o) - T\{c_v \ell n \frac{T}{T_o} + \underline{\underline{R}} \ell n \frac{v}{v_o} + s_o\}$

(b) $f(T,M) = c_M(T - T_o) - T\{c_M \ell n \frac{T}{T_o} - \frac{M^2 - M_o^2}{2C} + s_o\}$

(c) $f(T,\varepsilon) = \frac{1}{2} E\varepsilon^2 + \alpha E\varepsilon T_o + c_\varepsilon(T - T_o) - T(c_\varepsilon \ell n \frac{T}{T_o} + \alpha E\varepsilon + s_o)$.

Note that all expressions contain the troublesome term $-Ts_o$. This can be properly normalized only with the aid of the Third Law of thermodynamics. In the case of an incompressible fluid x = const, so that all derivatives $\partial .../\partial x$ become singular. This does not occur when derivatives with respect to P are taken. Thus,

$$g(P,T) = h(P,T) - Ts(P,T) = v(P - P_o) + c(T - T_o) - T(c\ell n \frac{T}{T_o} + s_o) .$$

Solution to Problem 1.5

Each perfect-gas component in a mixture contributes the term

$$-n_i \underline{\underline{R}} \ln x_i > 0$$

to the entropy of the mixture. Hence

$$S = ns = \sum_i n_i s_i(P,T) - \sum_i n_i \underline{\underline{R}} \ln x_i$$

with

$$s_i(P,T) = c_{pi} \ln \frac{T}{T_o} - \underline{\underline{R}} \ln \frac{P}{P_o} + s_{oi}$$

Thus, the simplest fundamental equation is that where $h = h(s,P)$, with

$$H = nh = \sum_i n_i h_i = \sum n_i c_{pi} (T - T_o)$$

It follows that

$$h = T_o\left(\frac{T}{T_o} - 1\right) \sum_i x_i c_{pi} \tag{a}$$

and that

$$s - s_o = \left(\ln \frac{T}{T_o}\right) \sum_i x_i c_{pi} - \left\{\underline{\underline{R}} \ln\left(\frac{P}{P_o}\right)\right\} \sum_i x_i - \underline{\underline{R}} \sum_i x_i \ln x_i \tag{b}$$

or

$$\frac{s - s_o}{\underline{\underline{R}}} = \left(\ln \frac{T}{T_o}\right) \sum_i x_i \frac{c_{pi}}{\underline{\underline{R}}} - \sum_i \ln x_i^{x_i} - \ln\left(\frac{P}{P_o}\right)$$

Note that

$$\frac{c_{pi}}{\underline{\underline{R}}} = \frac{c_{pi}}{c_{pi} - c_{vi}} = \frac{\gamma_i}{\gamma_i - 1}$$

and put

$$\Gamma(x_i) = \sum_i x_i \frac{c_{pi}}{\underline{\underline{R}}} = \sum_i \frac{\gamma_i x_i}{\gamma_i - 1} .$$

Thus,

Problem 1.5 (continued)

$$\frac{s - s_o}{\underline{\underline{R}}\,\Gamma} = \ell n \frac{T}{T_o} - \sum_i \ell n\, x_i^{x_i/\Gamma} - \ell n\left(\frac{P}{P_o}\right)^{1/\Gamma}$$

or

$$\frac{T}{T_o} = \left(\frac{P}{P_o}\right)^{1/\Gamma}\left(\prod_i x_i^{x_i/\Gamma}\right) \exp\left[(s - s_o)/\underline{\underline{R}}\,\Gamma\right]$$

and

$$h(P,s) = \bar{c}_p T_o \left\{\left(\frac{P}{P_o}\right)^{1/\Gamma}\left(\prod_i x_i^{x_i/\Gamma}\right)\left(\exp(s - s_o)/\underline{\underline{R}}\,\Gamma\right) - 1\right\}$$

where

$$\bar{c}_p = \sum_i x_i\, c_{pi} \ .$$

It is now easy to specialize for two components, with $x_1 = x$ and $x_2 = 1 - x$. In this manner

$$\bar{c}_p = x\, c_{p1} + (1 - x)\, c_{p2}$$

$$\Gamma(x) = \frac{x\gamma_1}{\gamma_1 - 1} + \frac{(1 - x)\gamma_2}{\gamma_2 - 1}$$

and

$$h(P,s,x) = \bar{c}_p\, T_o \left\{\left(\frac{P}{P_o}\right)^{1/\Gamma(x)}\left[x^{x/\Gamma(x)}(1 - x)^{(1-x)/\Gamma(x)} \exp\left((s - s)_o/\underline{\underline{R}}\Gamma(x)\right)\right] - 1\right\}.$$

In order to obtain the form $u = u(s,v,x)$, we write

$$s(T,v,x) = \sum_i x_i\, c_{vi} \ell n \frac{T}{T_o} + \underline{\underline{R}}\, \ell n \frac{v}{v_o} - \sum_i x_i\, \underline{\underline{R}}\, \ell n x_i + s_o \qquad \text{(a')}$$

with

$$u = \sum x_i\, c_{vi}(T - T_o) \ , \qquad \text{(b')}$$

and proceed in an analogous fashion as before.

The forms $g(T,P,x)$ and $f(T,v,x)$ are obtained directly from equations (a), (b) or (a'), (b'), respectively

Solution to Problem 1.8

The Gibbs equation for this system is

$$dG = -SdT + VdP + \sum_i \mu_i \, dn_i \ .$$

In a chemically active system

$$\frac{dn_i}{\nu_i} = d\xi \quad (\nu_i > 0 \text{ for products; } \nu_i < 0 \text{ for reactants}). \tag{a}$$

Hence

$$dG = -SdT + VdP + d\xi \sum_i \mu_i \nu_i = -SdT + VdP - Ad\xi, \text{ where } A = -\sum_i \mu_i \nu_i \ .$$

Since $dG = -dW^o$ at $T = \text{const}$, $P = \text{const}$, we find $dW^o = Ad\xi$. When the process is performed irreversibly $(dW^o = 0)$, the rate of entropy production is

$$\dot{\theta} = \frac{A\dot{\xi}}{T}$$

Evidently $\dot{\theta} = 0$ (or $dG = 0$) for

$$A = 0 \quad \text{or} \quad \sum \mu_i \nu_i = 0$$

Integrating the definition (a) from $n_i = n_i^o$ to n_i , we find

$$n_i = n_i^o + \nu_i \xi$$

$$n = \sum_i n_i = \sum_i n_i^o + \xi \sum_i \nu_i \ .$$

$$x_i = \frac{n_i^o + \nu_i \xi}{\sum_i n_i^o + \xi \sum_i \nu_i} \ . \tag{b}$$

On the other hand

Problem 1.8 (continued)

$$\mu_i = g_i(P,T) + \underline{\underline{R}}\,T\,\ell n\,x_i = g_i^o(T) + \underline{\underline{R}}\,T\,\ell n\,\frac{P}{P^*} + \underline{\underline{R}}\,T\,\ell n\,x_i\ ,$$

so that, either

$$\sum_i \nu_i\,g_i(P,T) + \sum_i \nu_i\,\underline{\underline{R}}\,T\,\ell n\,x_i = 0$$

or

$$\sum_i \nu_i g_i^o(P^*,T) + \sum_i \nu_i\,\underline{\underline{R}}T\,\ell n\,\frac{P}{P^*} + \sum_i \nu_i \underline{\underline{R}}T\,\ell n\,x_i = 0\ .$$

The two forms yield:

$$\prod_i x_i^{\nu_i} = \exp(-\Delta G/\underline{\underline{R}}T)$$

or

$$\prod_i x_i^{\nu_i} = \left(\frac{P}{P^*}\right)^{-\Delta\nu} \exp(-\Delta G^o/\underline{\underline{R}}T)$$

where

$$\Delta G = \sum_i \nu_i g_i(P,T);\quad \Delta G^o = \sum_i \nu_i g_i^o;\quad \Delta\nu = \sum_i \nu_i\ .$$

Combining with (a) we obtain

$$\prod_i \left\{\frac{n_i^o + \nu_i\xi}{\sum_i n_i^o + \xi\sum_i \nu_i}\right\}^{\nu_i} = \exp(-\Delta G/\underline{\underline{R}}T)$$

$$= \left(\frac{P}{P^*}\right)^{-\Delta\nu} \exp(-\Delta G^o/\underline{\underline{R}}T)\ .$$

Solution to Problem 1.9

The only additional complications consist in two facts:

First, we must write

$$dn_i = \sum_j \nu_{ij} \, d\xi_j$$

for each of the r reactions j and integrate each term from 0 to ξ_j :

$$n_i = n_i^o + \sum_j \nu_{ij} \, \xi_j \quad .$$

Secondly, there are now r equations leading to an equal number of equilibrium conditions:

$$A_j = -\sum \nu_{ij} \mu_i = 0 \ .$$

Here we note that the chemical potential μ_i for every species must be computed from

$$x_i = \frac{n_i^o + \sum_j \nu_{ij} \xi_j}{\sum_i n_i^o + \sum_i \sum_j \nu_{ij} \xi_j} \quad .$$

We are led to r coupled equations of the type considered in Problem 1.8.

Solution to Problem 1.10

(a) The state in each compartment is described by two independent properties, say, T_a, P_a and T_b, P_b . Since there is no interaction between the compartments, the total number of independent variables is $f = 4$.

(b) Now, the total number has decreased to $f = 2$.

The final state can be determined from the energy equation

$$m_a c_v T_a + m_b c_v T_b = (m_a + m_b) c_v T_2 \ .$$

$$T_2 = \frac{m_a T_a + m_b T_b}{m_a + m_b}$$

independently of the pressures (perfect gas) $P_a = m_a \underline{R} T_a / V_a$ and $P_b = m_b \underline{R} T_b / V_a$. The final pressure

$$P_2 = (m_a + m_b) \underline{R} T_2 / (V_1 + V_2)$$

$$= \frac{(m_a T_a + m_b T_b)\ \underline{R}}{V_1 + V_2} \ .$$

(c) Since the gas is in motion, its state is described by three independent fields. We can choose these as

$$s(\tilde{r},t),\ u(\tilde{r},t),\ \tilde{v}(\tilde{r},t)$$

Thus, at any instant

$$E(t) = \int_V \rho(\tilde{r},t) \left\{ u(\tilde{r},t) + \frac{1}{2} \tilde{v}^2(\tilde{r},t) \right\} dV$$

$$S(t) = \int_V \rho(\tilde{r},t) s(\tilde{r},t) dV \ .$$

In a concrete case, the fields must be determined by solving the following set of differential equations:

Problem 1.10 (continued)

1 equation of continuity

3 equations of momentum conservation

1 energy equation

1 fundamental equation of state

subject to the given initial and boundary conditions.

Solution to Problem 1.11

(a) $f = 4$

(b) $f = 2$ (the mol fraction $x = n_a/(n_a + n_b)$ which characterizes the composition is not an independent variable, because it is fixed by the initial masses m_a, m_b.

The final state follows from

$$m_a c_{va} T_a + m_b c_{vb} T_b = (m_a + m_b) \frac{m_a c_{va} + m_b c_{vb}}{(m_a + m_b)} T_2$$

$$T_2 = \frac{m_a c_{va} T_a + m_b c_{vb} T_b}{m_a c_{va} + m_b c_{vb}}$$

$$P_2 = P_{2a} + P_{2b} = \frac{m_a R_a T_2}{V_a + V_b} + \frac{m_b R_b T_2}{V_a + V_b} \quad ,$$

when P_{2a}, P_{2b} are the partial pressures of the two gasses in state 2.

Hence

$$P_2 = \frac{m_a R_a + m_b R_b}{V_a + V_b} T_2$$

(c) The additional complication arises from the fact that now we must add an equation describing the process of diffusion in the system. This adds one more field $x(\underset{\sim}{r},t)$ which has to be determined. The equations for $E(t)$ and $S(t)$ remain formally the same as in Problem 1.10, but it must not be forgotten that now $u = u(T,P,x)$ and $s = s(T,P,x)$. In a perfect gas $u = u(T,x)$ only.

Solution to Problem 1.12

We assume that compartment a contains n_a mols of gas A_m whereas compartment b contains n_b mols of gas B_n, but that the composition is not necessarily stoichiometric $(n_a/n_b \neq a/b)$.

(a) $f = 4$

(b) $f = 2$.

Even though a new species $A_{am}B_{bn}$ appears, its quantity is completely determined by the extent of the reaction, ξ, whose value is completely fixed when, say, the temperature and pressure in state 2 are prescribed. (The fact that $f = 2$ is known as Duhem's theorem; see J. Kestin, *A Course in Thermodynamics*, vol. 2, Blaisdell 1968, p. 345). If the reaction could be constrained internally (anticatalyst), we would have $f = 3$.

In order to simplify the task, we shall assume that compartment a contains $n_a = m_a/2$ mols of H_2 and that compartment b contains $n_b = \frac{1}{2} n_a = m_b/32$ mols O_2 (stoichiometric ratio). The extension to arbitrary quantities $(n_b \neq \frac{1}{2} n_a)$ presents no essential difficulties. We assume, further, that in state 2 the chemical equilibrium is unconstrained, denoting the extent of the reaction by ξ_2. Then in state 2

$$n_{H_2O} = n^o \xi_2 , \qquad x_{H_2O} = \frac{2\xi_2}{3 - \xi_2}$$

$$n_{H_2} = n^o(1 - \xi_2), \quad x_{H_2} = \frac{2(1 - \xi_2)}{3 - \xi_2}$$

$$n_{O_2} = n^o \frac{1}{2} (1 - \xi_2), \quad x_{O_2} = \frac{1 - \xi_2}{3 - \xi_2}$$

$$n = n^o(3 - \xi_2)/2 \qquad \text{where } n^o = n_a$$

The law of mass action is

Problem 1.12 (continued)

$$\frac{\xi_2(3-\xi_2)^{1/2}}{(1-\xi_2)^{3/2}} = \left(\frac{P_2}{P^*}\right)^{-1/2} K_p(T_2) \quad , \tag{a}$$

where $K_p(T)$ is a tabulated function, P^* is the standard pressure, and P_2 and T_2 are still unknown.

The perfect-gas law for state 2 yields

$$P_2 = \frac{\frac{1}{2}n^o(3-\xi_2)}{V_a + V_b} \underline{\underline{R}}\, T_2 \; . \tag{b}$$

The third equation for P_2, T_2, ξ_2 is provided by the energy equation

$$U_2 = U_{a1} + U_{b1} \; , \tag{c}$$

where

$$U_{a1} = n_a u^o_{H_2}(T_a) \qquad U_{b1} = \frac{1}{2} n_a u^o_{O_2}(T_b) \; , \tag{c_1, c_2}$$

and $u^o_i(T)$ are tabulated functions; if only a tabulation of $h^o_i(T)$ is available, we compute $u^o_i(T) = h^o_i(T) - Pv = h^o_i(T) - \underline{\underline{R}}T$.

The energy at state 2 is:

$$U_2 = n^o[u^o_{H_2}(T_2) + \frac{1}{2} u^o_{O_2}(T_2)] + n^o\xi_2 \Delta U^o(T) \; , \tag{c_3}$$

where $\Delta U^o(T) = \Delta H^o(T) - P\Delta V = \Delta H^o(T) - \underline{\underline{R}}T\Sigma\nu_i = \Delta H^o(T) + \frac{1}{2}\underline{\underline{R}}T$ is the energy of the reaction.

Substituting (c_1, c_2, c_3) into (c), we obtain three equations: (a), (b), (c) which determine the state. Numerical calculations can be performed on a computer.

If the reader has difficulties in understanding equation (c_3), he may remember that the energy (or enthalpy) of a reaction must be used to normalize

Problem 1.12 (continued)

the tabulated functions $U_p^o = \Sigma n_i u_i^o$ for the products and $U_r^o = \Sigma n_i u_i^o$ for the reactants. Denoting the normalized functions by $\mathcal{U}_p(T)$ and $\mathcal{U}_r^o(T)$, we note that

$$\mathcal{U}_p^o(T) - \mathcal{U}_r^o(T) = \Delta U^o(T) \ . \qquad \text{(tabulated or derived from } \Delta H^o(T))$$

We are free to put $\mathcal{U}_r^o(T) = U_r^o(T)$, so that

$$\mathcal{U}_p^o(T) = U_r^o(T) + \Delta U^o(T) \ .$$

On the other hand, for the equilibrium mixture

$$\mathcal{U}_{mix} = \xi \mathcal{U}_p^o + (1 - \xi)\, \mathcal{U}_r^o = \xi(U_r^o + \Delta U^o) + (1 - \xi) U_r^o = U_r^o + \xi \Delta U^o \ ,$$

which is equation (c_3) .

(c) The equations for $E(t)$ and $S(t)$ are again formally the same as in Problem 1.10; it must be remembered that now $u = u(T,P,\xi)$ and $s = s(T,P,\xi)$. Evidently, the field equations become even more complex.

Solution to Problem 1.13

Since the metal is incompressible, and since, by implication its stress is assumed to be identically equal to zero, we need only one field, namely $T(x)$.

Since energy is conserved, we have

$$\int_o^\ell c[T(x) - T_2]A\,dx = A\ell c(T_f - T_2)$$

since $u = c(T - T_2)$. Hence

$$T_f = T_2 + \frac{1}{2}(T_1 - T_2) \ .$$

Noting that $s = c\ \ell n\left(\frac{T}{T_2}\right)$ and that

$$T(x)/T_2 = (T_1/T_2 - 1)\ (x/\ell) + 1$$

We calculate

$$\Delta S = \rho A\ell c\ \ell n\ \frac{T_f}{T_2} - \int_o^\ell c\ \ell n\ \frac{T(x)}{T_2}\ A\rho dx$$

$$\int \ell n(bx + 1)dx = \frac{bx+1}{b}\ \ell n(bx + 1) - \frac{bx+1}{b} \quad \text{where} \quad b = (T_1/T_2 - 1)/\ell$$

$$\int_o^\ell \ell n(bx + 1)dx = \ell\left(\frac{T_1/T_2}{T_1/T_2 - 1}\ \ell n\ \frac{T_1}{T_2} - 1\right)$$

Hence

$$S = \rho A\ell c\left\{\ell n\ \frac{1}{2}\left(\frac{T_1}{T_2} + 1\right) - \frac{T_1/T_2}{T_1/T_2 - 1}\ \ell n\ \frac{T_1}{T_2} + 1\right\} \ .$$

Solution to Problem 1.14

The maximum work at V = const is

$$W_{max} = W^o = U_1 - U_2 - T_o(S_1 - S_2) .$$

In this case $U_1 - U_2 = 0$ and

$$W_{max} = T_o \Delta S$$

i.e.

$$W_{max} = \rho A \ell c \; T_o \left\{ \ell n \frac{1}{2}\left(\frac{T_1}{T_2} + 1\right) - \frac{T_1/T_2}{T_1/T_2 - 1} \ell n \frac{T_1}{T_2} + 1 \right\}$$

Solution of Problem 2-2

We may employ the representation of x and y in polar coordinates as

$$x = r \cos \theta$$

$$y = r \sin \theta \quad .$$

This leads to expressions for the time derivatives $\dot{x}$, and $\dot{y}$,

$$\dot{x} = \dot{r} \cos \theta - r \dot{\theta} \sin \theta$$

$$\dot{y} = \dot{r} \sin \theta + r \dot{\theta} \cos \theta \quad .$$

The kinetic energy is then

$$\frac{1}{2} m(\dot{x}^2 + \dot{y}^2) = \frac{1}{2} m(\dot{r}^2 + r^2\dot{\theta}^2) \quad .$$

The total energy is

$$\frac{1}{2} m(\dot{x}^2 + \dot{y}^2) + \frac{\alpha}{2}(x^2 + y^2) = \frac{1}{2} m(\dot{r}^2 + r^2\dot{\theta}^2) + \alpha r^2/2 \quad .$$

The term $\alpha r^2/2$ is the potential energy.

The term $\frac{1}{2} m\dot{r}^2 \equiv P_r^2/2m$ where P_r is the component of the momentum in the radial direction.

The term $mr^2\dot{\theta}^2/2$ may be identified with the energy of a particle of mass m moving along a circle of radius r with angular velocity $\dot{\theta}$. The angular momentum of such a system is $mr^2\dot{\theta} \equiv P_\theta$, and with this definition $\frac{1}{2} mr^2\dot{\theta}^2 = P_\theta^2/2mr^2$. The total energy may be written

$$\frac{1}{2m} \left(P_r^2 + P_\theta^2/r^2\right) + \alpha r^2/2 \quad .$$

Solution to Problem 2-12

a) Write $r = |\underline{\underline{r}}|$, where $\underline{\underline{r}} = x\,\hat{i} + y\,\hat{j} + z\,\hat{k}$. Here $\hat{i}$, $\hat{j}$, $\hat{k}$ are unit vectors in the x , y , z directions respectively. To compute the force we need to evaluate $\nabla\phi(r)$. This can be most easily evaluated by noticing the relation

$$\nabla\phi(r) = (\hat{i}\,\frac{\partial r}{\partial x} + \hat{j}\,\frac{\partial r}{\partial y} + \hat{k}\,\frac{\partial r}{\partial z})\ \frac{\partial\phi}{\partial r} = \frac{\underline{\underline{r}}}{r}\,\frac{\partial\phi}{\partial r}$$

where

$$\frac{\partial\phi}{\partial r} = \frac{-2c}{r_o}\ \varepsilon\ \{\exp\ (-\ 2\ c(r-d)/r_o) - \exp\ (-\ c(r-d)/r_o)\}\ .$$

The force $\underline{\underline{F}}(r)$, is then

$$\underline{\underline{F}} = -\ \nabla\phi = \frac{2c\varepsilon}{r_o}\ \{\exp\ (-2c(r-d)/r_o) - \exp\ (-c(r-d)/r_o)\}\ \frac{\underline{\underline{r}}}{r}\ .$$

It is clear from the above expression that $\underline{\underline{F}} = 0$ at $r = r_m = d$ and that $\phi(r)$ has an extremum at r_m .

b) By substitution, we may easily verify that $\phi(r = d) = -\ \varepsilon$. To prove that ϕ actually possesses a minimum at $r = d$, we may employ the criterion that $\partial^2\phi/\partial r^2 > 0$ at $r = r_m$ for the function to assume a minimum.

c) The distance r_o locates the value of r for which the potential is zero. This implies that

$$0 = \exp\ (-2c(r_o-d)/r_o) - 2\ \exp\ (-c(r_o-d)/r_o)\ ,$$

or

$$-\ 2c(r_o-d)/r_o = \ln 2 - c(r_o-d)/r_o\ .$$

This equation leads to the result that

$$r_m = d = r_o(1 + \frac{\ln 2}{c})\ .$$

Solution to Problem 2-16

For a system of n molecules there are 3n degrees of freedom. Three of these correspond to the location of the center of mass $(f_{tr} = 3)$.

This leaves $3n - 3$ internal degrees of freedom

a) Diatomic molecule:

In this case $n = 2$ and $3n - 3 = 3$

Here $f_v = 1$,

so $f_r = 2$

There is no rotation about the axis of the molecule.

b) Linear, homonuclear triatomic molecule:

In this case $n = 3$ and $3n - 3 = 6$

Here $f_r = 2$,

since there is no rotation about the axis of the molecule. Consequently

$$f_v = 4$$

Two of these vibrational modes are parallel to the molecule axis, and two are perpendicular to the axis.

c) Nonlinear, homonuclear triatomic molecule:

Here, again, there are 6 internal degrees of freedom, but three of these are rotational, $f_r = 3$. Consequently $f_v = 3$. It is interesting to note that the nonlinear triatomic molecule has one less vibrational degree of freedom than a linear triatomic molecule.

If the molecules are heteronuclear the numbers of degrees of freedom given above are unchanged.

Solution to Problem 2-19

a) The volume per molecule, v , is given by

$$v = V/N = m/\rho = (43 \times 10^{-24}/1.65)\ \text{cm}^3 = 26.06 \times 10^{-24}\ \text{cm}^3 \ .$$

The simple estimate for the distance between argon atoms in this solid is $v^{1/3} = 2.96 \times 10^{-8}$ cm.

b) If we assume that argon atoms are hard spheres of diameter σ , then we may obtain σ as

$$v = \sigma^3/\sqrt{2} = 26.06 \times 10^{-24}\ \text{cm}^3 \ ,$$

or

$$\sigma^3 = 36.85 \times 10^{-24}\ \text{cm}^3 \ ,$$

or

$$\sigma = 3.33 \times 10^{-8}\ \text{cm} \ .$$

Solution to Problem 3-7

a) Employing the de Broglie relation we find that the wavelength of the electron is

$$\lambda = \underline{\underline{h}}/p = \underline{\underline{h}}/mv = \frac{6.6254 \times 10^{-27} \text{ erg-sec}}{(3 \times 10^{6} \text{ cm/sec})\ (9.11 \times 10^{-28} \text{ gm})} .$$

$$= 2.42 \times 10^{-6} \text{ cm}.$$

b) The wavelength of the projectile is

$$\lambda = \underline{\underline{h}}/p = \frac{6.6254 \times 10^{-27}}{12 \times 6 \times 10^{4}} \text{ cm} = 9.2 \times 10^{-33} \text{ cm} .$$

The wavelength of the projectile is so small that diffraction effects are undetectable, for in order to detect such effects one needs a diffracting device with a slit size of the order of 10^{-32} cm. No such device exists.

Solution to Problem 3.11

a) By inspection we see that any solution of the form

$$u(x,t) = A \sin \frac{n\pi x}{\ell} \cos \frac{n\pi a}{\ell} t \quad ,$$

where n is an integer, satisfies the given boundary and initial conditions.

b) The suggested substitution for a is in error. Instead, one should replace a by the phase velocity $\nu\lambda$, and use the relation $E = \underline{\underline{h}}\nu$. Hence, we write

$$\Psi(x,t) = ce^{i(2\pi/\lambda)(x-\nu\lambda t)} = ce^{2\pi i(px-Et)/\underline{\underline{h}}} \quad .$$

This is a solution of the equation for the vibrating string equation, but it also satisfies the equation

$$\frac{i\underline{\underline{h}}}{2\pi} \frac{\partial\Psi}{\partial t} = \frac{-\underline{\underline{h}}^2}{8m\pi^2} \frac{\partial^2\Psi}{\partial x^2} \quad ,$$

which is the time-dependent Schrodinger equation for a free particle.

Solution to Problem 3-13

a) We determine a by inserting this wave function into the Schrödinger equation

$$\frac{-\underline{\underline{h}}^2}{8\pi^2 m}\frac{\partial^2}{\partial x^2} Ae^{-x^2/2a} + \frac{1}{2} k\, x^2 Ae^{-x^2/2a} = EAe^{-x^2/2a} .$$

This leads to the result

$$-\left(\frac{\underline{\underline{h}}^2}{8\pi^2 m}\right)\left(\frac{x^2}{a^2} - \frac{1}{a}\right) + \frac{1}{2} kx^2 = E .$$

Comparing powers of x we find that

$$E = \frac{\underline{\underline{h}}^2}{8\pi^2 ma}$$

and that $\underline{\underline{h}}^2/8\pi^2 ma^2 = \frac{1}{2} k$

or

$$a = (\underline{\underline{h}}^2/4\pi^2 mk)^{1/2} = \frac{\underline{\underline{h}}}{2\pi} (mk)^{-1/2}$$

We determine the normalizing factor A by the requirement that

$$\int_{-\infty}^{\infty} |\psi|^2 \, dx = AA^* \int_{-\infty}^{\infty} e^{-x^2/a} \, dx = AA^* \sqrt{\pi a} = 1 .$$

Therefore $|A| = (\pi a)^{-1/4}$ and we may take A to be real, for convenience. Moreover

b) The probability of observing the particle at point x is identically zero, but the probability of observing the particle in the range x to x + dx is

$$|\psi(x)|^2 \, dx = (\pi a)^{-1/2} e^{-x^2/a} \, dx$$

Solution to Problem 3-13 continued

c) $\langle x \rangle = \int_{-\infty}^{\infty} x|\psi|^2 dx = (\pi a)^{-1/2} \int_{-\infty}^{\infty} x e^{-x^2/a} dx = 0$.

d) $\langle x^2 \rangle = \int_{-\infty}^{\infty} \psi^* x^2 \psi \, dx = (\pi a)^{-1/2} \int_{-\infty}^{\infty} x^2 e^{-x^2/a} dx = a/2$.

e)
$$\langle p^2/2m \rangle = -\frac{h^2}{8m\pi^2} \int_{-\infty}^{\infty} \psi^* \frac{\partial^2}{\partial x^2} \psi \, dx$$
$$= -\frac{h^2}{8m\pi^2} (\pi a)^{-1/2} \int_{-\infty}^{\infty} e^{-x^2/2a} \frac{\partial^2}{\partial x^2} e^{-x^2/2a} dx$$
$$= -\frac{h^2}{8m\pi^2} (\pi a)^{-1/2} \int_{-\infty}^{\infty} e^{-x^2/a} \left(\frac{x^2}{a^2} - \frac{1}{a}\right) dx$$
$$= h^2/16ma\pi^2 \, .$$

f) $\langle p^2/2m + kx^2/2 \rangle = h^2/16ma\pi^2 + ka/4$.

Using the result from part a) that

$$a = \frac{h}{2\pi} (mk)^{-1/2} \, .$$

We find that

$$\langle p^2/2m + kx^2/2 \rangle = \frac{h}{4\pi} \left(\frac{k}{m}\right)^{1/2} = E \, .$$

Notice further that $E = \frac{1}{2} h \nu$, where ν is the classical frequency of the harmonic oscillator.

Solution to Problem 3-20

a) To verify that the wave function is normalized we need to compute the integral

$$\int \psi^* (\underline{r})\, \psi(\underline{r})\, d\underline{r} = \int_0^\infty \int_0^\pi \int_0^{2\pi} |\psi|^2\, r^2 \sin\theta\, dr\, d\theta\, d\phi$$

$$= \frac{4\pi}{\pi a^3} \int_0^\infty r^2 e^{-2r/a}\, dr = \frac{4}{a^3} \frac{2}{(2/a)^3} = 1 \,.$$

b) The probability of finding the electron in a small shell of radius Δr about the origin is given by the integral

$$\int_0^{\Delta r} \int_0^\pi \int_0^{2\pi} |\psi(\underline{r})|^2\, r^2 \sin\theta\, dr\, d\theta\, d\phi = \frac{4\pi}{\pi a^3} \int_0^{\Delta r} r^2 e^{-2r/a}\, dr \simeq \frac{4}{3} \frac{(\Delta r)^3}{a^3} \,.$$

c) The probability that the particle is within the distance a from the origin is given by the integral

$$\int_0^a \int_0^\pi \int_0^{2\pi} |\psi(r)|^2\, r^2 \sin\theta\, dr\, d\theta\, d\phi$$

$$= \frac{4}{a^3} \int_0^a r^2 e^{-2r/a}\, dr = 1 - \frac{5}{e^2} \,.$$

Solution to Problem 4-10

Let E_1 denote the event that the bill becomes law.

Let E_2 denote the event that Congress passes the bill.

Let E_3 denote the event that the President does not veto the bill, if passed by Congress.

Then $\pi(E_2) = 1/3$; $\pi(E_3) = 1/2$

and $\pi(E_1) = \pi(E_2)\ \pi(E_3) = (1/3)\ (1/2) = 1/6$.

Solution to Problem 4-14

If we consider the function $F(x_1, \dots x_i, \dots)$ defined by

$$F(x_1, \dots , x_i \dots) = - \sum x_i \ln x_i + \lambda \sum x_i \quad .$$

The extremum condition becomes

$$\frac{\partial F}{\partial x_i} = 0 \quad \text{for all} \quad i$$

or

$$- 1 - \ln x_i^o + \lambda = 0$$

which yields

$$\ln x_i^o = (\lambda - 1) \quad \text{for all} \quad x_i \ .$$

λ is determined from the relation that

$$1 = \sum x_i^o = \sum_i \exp (\lambda - 1) \ .$$

To see that this is a maximum we observe that

$$\left. \frac{\partial^2 F}{\partial x_i^2} \right|_{x_i = x_i^o} = - \frac{1}{x_i^o} < 0$$

which implies that $- \sum x_i \ln x_i$ achieves its maximum at $x_i = x_i^o$.

Solution to Problem 4-16

This problem is similar in form to radioactive decay. We divide the time interval 0 to T into intervals of length δt such that $N\delta t = T$, and δt is small compared to T.

The probability that no defect appears in any one of these intervals is $(1 - \lambda\delta t) = (1 - \lambda T/N)$.

The probability that no defect appears in time T is therefore $P = (1 - \lambda T/N)^N$.

For large N, a suitable approximation for P is

$$P = e^{-\lambda T} .$$

Solution to Problem 4-30

The volume of an N-dimensional sphere of radius R may be formulated as the multiple integral

$$V_N(R) = \underset{\sum x_i^2 \leq R^2}{\int dx_1 \cdots \int dx_N}$$

The substitution $x_i = Rx_i'$ leads to

$$V_N(R) = R^N \underset{\sum x_i'^2 \leq 1}{\int dx_1' \cdots \int dx_n'} = R^N V_N(1)$$

where $V_N(1)$ is the volume of the N-dimensional unit sphere.

The corresponding surface areas may be most easily obtained by noticing that in any number of dimensions, the volume, $dV_N(R)$ of a spherical shell of thickness dR is given by

$$dV_N(R) = S_N(R)\ dR$$

Where $S_N(R)$ is the area of an N-dimensional sphere of radius R .

Consequently,

$$S_N(R) = dV_N(R)/dR = NR^{N-1}\ V(1)$$

Therefore

$$S_N(1) = NV(1)$$

and

$$S_N(R) = R^{N-1}\ S_N(1)$$

Solution to Problem 5.1

Suppose two trajectories in Γ-space intersect at some point $(\underline{\underline{r}}_1, \underline{\underline{p}}_1, \underline{\underline{r}}_2, \underline{\underline{p}}_2, \ldots \underline{\underline{p}}_n)$. Given these values of the positions and momenta of a system of particles, both the forward and the time reversed motions are uniquely specified. Therefore, if two trajectories have one point in common, they must actually coincide.

Solution to Problem 5-5

Let N_1 be the number of particles with energy $E_1 = 0$ and N_2 be the number with energy E_2 . Since the total energy of the system is given to be E , we must have the two equations

$$N_1 + N_2 = N$$

$$N_2 E_2 = E \quad .$$

The number of microstates is then given by

$$\Omega(E) = \frac{N!}{N_1 ! \, N_2 !} = \frac{N!}{(\frac{E}{E_2}) ! \, (N - \frac{E}{E_2}) !} \quad .$$

a) The entropy S is

$$S = \underline{\underline{k}} \, \ell n \, \Omega(E) = \underline{\underline{k}} \, \ell n \, N ! - \underline{\underline{k}} \, \ell n \, (\frac{E}{E_2}) ! - \underline{\underline{k}} \, \ell n \, (N - \frac{E}{E_2}) ! \quad .$$

The temperature is determined through the thermodynamic relation

$$T = \left(\frac{\partial E}{\partial S}\right)_N = \left(\frac{\partial S}{\partial E}\right)_N^{-1} = \frac{E_2}{\underline{\underline{k}}} \left[\ell n \left(\frac{NE_2 - E}{E}\right)\right]^{-1}$$

where Stirling's formula has been used.

b) The entropy will be a maximum when $N_1 = N_2 = N/2$.
The corresponding energy is $NE_2/2$.

Solution of Problem 5-11

Since each of the systems in the ensemble consists of a single free particle, Liouville's equation takes the form

$$\frac{\partial}{\partial t} \rho(\underline{\underline{r}}, \underline{\underline{p}}, t) + (v_x \frac{\partial}{\partial x} + v_y \frac{\partial}{\partial y} + v_z \frac{\partial}{\partial z}) \rho(\underline{\underline{r}}, \underline{\underline{p}}, t) = 0$$

since $\dot{p}_x = \dot{p}_y = \dot{p}_z = 0$.

The solution of this equation is:

$$\rho(\underline{\underline{r}}, \underline{\underline{p}}, t) = \rho(\underline{\underline{r}} - \frac{\underline{\underline{p}}}{m} t, \underline{\underline{p}}, 0) ,$$

that is to say, the value of the distribution function at the point $\underline{\underline{r}}, \underline{\underline{p}}$ at the time t, is equal to the value of the distribution function at $t = 0$, evaluated at the point $\underline{\underline{r}} - \underline{\underline{v}}\, t$. In this case, the motion along a trajectory in Γ-space corresponds to the free motion of a single particle.

Solution to Problem 6-9

This is a one-dimensional problem with

$$\mathcal{H}(p,q) = p^2/2m + aq^2 + bq^3 + cq^4 \ .$$

The single-particle partition function Z, is

$$Z = \int_{-\infty}^{+\infty} \int \exp - \beta\mathcal{H}(p,q)\ dpdq$$

$$= \left(\frac{2\pi m}{\beta}\right)^{1/2} \int_{-\infty}^{\infty} \exp\ (-\ a\beta q^2 - b\beta q^3 - c\beta q^4)\ dq \ .$$

If we change the variable of integration from q to $(a\beta)^{1/2}\ q$, we can write Z as

$$Z = \left(\frac{2\pi m}{a\beta^2}\right)^{1/2} \int_{-\infty}^{\infty} \exp\left[-\ q^2 - (b/a^{3/2}\beta^{1/2})q^3 - (c/a^2\beta)q^4\right] dq \ .$$

We now expand the integrand in powers of b and c to obtain

$$Z = \left(\frac{2\pi m}{a\beta^2}\right)^{1/2} \int_{-\infty}^{\infty} e^{-q^2} \left[1 - (b/a^{3/2}\beta^{1/2})\ q^3 + \ldots\right]\left[1 - (c/a^2\beta)q^4 + \ldots\right] dq.$$

Since

$$\int_{-\infty}^{\infty} q^n\ e^{-q^2}\ dq = 0 \qquad \text{if } n \text{ is odd} \ ,$$

it follows that

$$Z = \left(\frac{2\pi m}{a\beta^2}\right)^{1/2} \int_{-\infty}^{\infty} e^{-q^2}\ (1 - \eta q^4 + \tfrac{1}{2}\ \xi q^6 + \tfrac{1}{2}\ \eta^2\ q^8 - \tfrac{1}{2}\ \eta\xi q^{10} + \ldots)\ dq \ ,$$

where

$$\eta = c/a^2\beta$$

$$\xi = b^2/a^3\beta \ .$$

Problem 6-9 (continued)

Evaluating the integrals we find

$$Z = \left(\frac{2\pi^2 m}{a\beta^2}\right)^{1/2} \left(1 - \frac{3}{4}\eta + \frac{15}{16}\xi + \frac{105}{32}\eta^2 - \frac{945}{64}\xi\eta + \ldots\right) .$$

The energy per particle is given by

$$E = \frac{1}{\beta}\left(1 - \frac{3}{4}\eta + \frac{15}{16}\xi + 6\eta^2 - \frac{225}{8}\xi\eta - \frac{225}{256}\xi^2 + \ldots\right) .$$

The specific heat per particle, C_V , is

$$C_V = \underline{\underline{k}}\left(1 - \frac{3}{2}\eta + \frac{15}{8}\xi + 18\eta^2 - \frac{675}{8}\xi\eta - \frac{675}{256}\xi^2 + \ldots\right) .$$

The entropy per particle is given by

$$S = \underline{\underline{k}}\left[\ell n\left(\frac{2\pi^2 m}{a\beta^2}\right)^{1/2} + 1 - \frac{3}{2}\eta + \frac{15}{8}\xi + 9\eta^2 - \frac{675}{16}\xi\eta - \frac{675}{512}\xi^2 + \ldots\right] .$$

Solution to Problem 6-15

The partition function of the system is given by

$$Z = \left(\sum_{m=1} e^{-m\beta\varepsilon_0}\right)^N = \left(\frac{e^{-\beta\varepsilon_0}}{1-e^{-\beta\varepsilon_0}}\right)^N ,$$

or

$$\ln Z = N\left[-\beta\varepsilon_0 - \ln\left(1 - e^{-\beta\varepsilon_0}\right)\right] .$$

The energy of this system is given by

$$E = -\frac{\partial \ln Z}{\partial \beta} = N\varepsilon_0 \left\{1 + \frac{e^{-\beta\varepsilon_0}}{1-e^{-\beta\varepsilon_0}}\right\} . \qquad (1)$$

The entropy of this system is given by

$$S = \underline{\underline{k}} \frac{\partial}{\partial T} (T \ln Z)$$

$$= \underline{\underline{k}} \{\ln Z + \beta E\} ,$$

or

$$S/\underline{\underline{R}} = \frac{1}{\underline{\underline{N}}} \{\ln Z + \beta E\} . \qquad (2)$$

From equation (1) we obtain

$$e^{-\beta\varepsilon_0}/(1 - e^{-\beta\varepsilon_0}) = \frac{E}{N\varepsilon_0} - 1 . \qquad (3)$$

Hence

$$\ln Z = N \ln \left[e^{-\beta\varepsilon_0}/(1 - e^{-\beta\varepsilon_0})\right] = N \ln \left(\frac{E}{N\varepsilon_0} - 1\right) .$$

From equation (3) we obtain

$$e^{\beta\varepsilon_0} = \frac{E/N\varepsilon_0}{E/N\varepsilon_0 - 1} ,$$

Problem 6-15 (continued)

or

$$\beta = \frac{1}{\varepsilon_o} \ell n \; (E/N\varepsilon_o) - \frac{1}{\varepsilon_o} \ell n \; (E/N\varepsilon_o - 1) \; .$$

Finally equation (2) becomes

$$S/\underset{=}{R} = \frac{E}{N\varepsilon_o} \ell n \frac{E}{N\varepsilon_o} - \left(\frac{E}{N\varepsilon_o} - 1\right) \ell n \left(\frac{E}{N\varepsilon_o} - 1\right) \; .$$

Notice that

$$\frac{\partial S}{\partial E} = \frac{\underset{=}{R}}{N\varepsilon_o} \ell n \left[\frac{E/N\varepsilon_o}{E/N\varepsilon_o - 1}\right] > 0 \; , \; \text{since} \quad E > N\varepsilon_o \quad .$$

Therefore the entropy increases monotonically with energy.

Solution to Problem 6.43

In order to solve this Problem we first need to convert the units of some quantities into more useful units. In particular we may use the conversion factors

$$1 \text{ atm} = 1.013 \times 10^{5} \text{ N/m}^{2}$$

$$1 \text{ eV} = 1.602 \times 10^{-12} \text{ erg} = 1.602 \times 10^{-19} \text{ J} .$$

Consequently

$$\Theta_i = (13.6)\ (1.602 \times 10^{-19})/1.38 \times 10^{-23} \text{ K}$$

$$= (13.6)\ (1.16 \times 10^{4}) \text{ K} \simeq 15.8 \times 10^{4} \text{ K} .$$

The extent of ionization ϕ is given in Equations (6.79a) and (6.79b) as

$$\frac{\phi^2}{1-\phi^2} = C\,\frac{T^{5/2}}{P}\, e^{-\Theta_i/T} .$$

Inserting the appropriate numerical factors we find that

$$\frac{\phi^2}{1-\phi^2} = \frac{(6.28)^{3/2}\ (9.1 \times 10^{-31})^{3/2}\ (1.38 \times 10^{-23})^{5/2}}{(6.625 \times 10^{-34})^{3}} \cdot \frac{(10^{4})^{5/2}}{1.013\times 10^{3}} \exp(-\ 15.8),$$

where we have inserted $g_a = 2$ to account for the two spin states of the electron, and we have set $g_i = 1$, since the proton spin states are accounted for in the nuclear partition function which we have not included.
Therefore

$$\frac{\phi^2}{1-\phi^2} = \frac{[(6.28)\ (.91)\ (.138)]^{3/2}\ (.138)}{(6.625)^{3}}\ 10^{9-15.8 \log e}$$

$$\approx 4.5 \times 10^{-2} ,$$

and $\phi \approx 2.1 \times 10^{-1}$

Solution to Problem 7-1

The Mayer f function for the Sutherland potential is

$$f(r) = e^{-\varepsilon_2(r)/kT} - 1 = \begin{cases} -1 & r < \sigma \\ e^{\frac{c}{kT}\left(\frac{\sigma}{r}\right)^6} - 1 & r > \sigma \end{cases} .$$

The second virial coefficient is given by

$$B(T) = \frac{-N}{2} \int f(r)\, d\mathbf{r}$$

$$= -2\pi N \left\{ - \int_0^{\sigma} r^2\, dr + \int_{\sigma}^{\infty} \left(\exp\left[\frac{c}{kT} \left(\frac{\sigma}{r}\right)^6 \right] - 1 \right) r^2\, dr \right. .$$

If $c/kT \ll 1$, we may expand the integrand of the second integral to obtain

$$B(T) = 2\pi N \left\{ \frac{\sigma^3}{3} - \int_{\sigma}^{\infty} \left[\frac{c}{kT}\left(\frac{\sigma}{r}\right)^6 + \frac{1}{2}\left(\frac{c}{kT}\right)^2 \left(\frac{\sigma}{r}\right)^{12} + \ldots \right] r^2\, dr \right.$$

$$= \frac{2}{3}\pi N \sigma^3 \left\{ 1 - \frac{c}{kT} - \frac{1}{6}\left(\frac{c}{kT}\right)^2 + \ldots \right\} .$$

The equation of state is

$$P = nkT \left\{ 1 + \frac{2}{3}\pi n \sigma^3 \left(1 - \frac{c}{kT} - \frac{1}{6}\left(\frac{c}{kT}\right)^2 + \ldots \right) + \ldots \right\} .$$

The internal energy per particle u is given by

$$u = \frac{3}{2} kT - \frac{2}{3}\pi n \sigma^3 \left(c + \frac{1}{3} \frac{c^2}{kT} \right) + \ldots .$$

Problem 7-1 (continued)

Hence, the specific heat is

$$C_v = \frac{3}{2}\underline{\underline{k}} + \frac{2}{9}\,\frac{\pi n \sigma^3 c^2}{\underline{\underline{k}} T^2} + \ldots \ .$$

Solution to Problem 7-8

The second virial coefficient is given by

$$B(T) = \frac{-N}{2} \int f(r)\, d\mathbf{r} = -2N\pi \int_0^{\infty} (e^{-\varepsilon_2(r)/kT} - 1)\, r^2\, dr\ .$$

By partial integration we have the result that

$$B(T) = \frac{-2\pi N}{3kT} \int_0^{\infty} r^3 \frac{d\varepsilon_2(r)}{dr} e^{-\varepsilon_2(r)/kT}\, dr = \frac{-N}{6kT} I\ ,$$

assuming that

$$r^3(e^{-\varepsilon_2(r)/kT} - 1) \rightarrow 0 \qquad \text{as } r \rightarrow \infty \text{ and as } r \rightarrow 0\ .$$

Solution to Problem 7-15

The pair potential for a gas of hard sphere molecules, $\varepsilon_2(r)$ is infinite for $r < \sigma$ and zero otherwise. Therefore. the function $\Phi(r)$, defined as $\Phi(r) = e^{-\beta\varepsilon_2(r)}$ is independent of temperature.

Consequently, the configurational partition function

$$Q_N = \int d\underline{\underline{r}}_1 \cdots \int d\underline{\underline{r}}_N \, \Phi_{12} \, \Phi_{13} \cdots \Phi_{N-1,N}$$

is independent of temperature. The partition function of the system is

$$Z = \frac{1}{N!}\left(\frac{2\pi m\underline{\underline{k}}T}{\underline{\underline{h}}^2}\right)^{3N/2} Q_N(V) ,$$

or

$$\ell n\ Z = \frac{3N}{2}\ \ell n \frac{2\pi m\underline{\underline{k}}T}{\underline{\underline{h}}^2} - \ell n\ N! + \ell n\ Q_N(V) .$$

The pressure is then given by

$$P = \underline{\underline{k}}T\left(\frac{\partial\ \ell n\ Z}{\partial V}\right)_T = \underline{\underline{k}}T\ \frac{\partial\ \ell n\ Q_N(V)}{\partial V}$$

and varies linearly with temperature, since Q_N is a function of V alone.

Solution to Problem 8-1

a) We use the expression

$$\Omega_\nu^{B.E.} = \prod_i \frac{(n_i^\nu + g_i - 1)!}{n_i^\nu!\,(g_i-1)!} .$$

If $g_i = 1$ for all i

$$\Omega_\nu^{B.E.} = \prod_i \frac{n_i^\nu!}{n_i^\nu!} = 1 .$$

b) If $g_i >> 1$, we may write

$$\ln \Omega_\nu^{B.E.} = \sum_{i=1}^{k} [\ln (n_i^\nu + g_i - 1)! - \ln n_i^\nu! - \ln (g_i - 1)!] ,$$

or, applying Stirling's formula, we find that

$$\ln \Omega_\nu^{B.E.} = \sum_{i=1}^{k} \left\{ g_i \ln \frac{n_i^\nu + g_i}{g_i} + n_i^\nu \ln \frac{n_i^\nu + g_i}{n_i^\nu} \right\} .$$

c) The most probable distribution is that which maximizes $\Omega_\nu^{B.E.}$ subject to the constraints that

$$N = \sum n_i^\nu$$

and

$$E = \sum \varepsilon_i n_i^\nu$$

remain constant.

We seek those values of n_i^ν which satisfy

Problem 8-1 (continued)

$$\delta \ln \Omega_{\nu}^{B.E.} = \sum_{i=1}^{k} [\ln (n_i^{\nu} + g_i) - \ln n_i^{\nu}]\, \delta n_i^{\nu} = 0$$

$$\sum_{i=1}^{k} \delta n_i^{\nu} = 0$$

and

$$\sum_{i=1}^{k} \varepsilon_i\, \delta n_i^{\nu} = 0 .$$

Using the undetermined multipliers α , and β , we obtain

$$\sum_{i=1}^{k} \left[\ln \frac{n_i^{\nu} + g_i}{n_i^{\nu}} - \alpha - \beta\varepsilon_i\right] \delta n_i^{\nu} = 0 .$$

Consequently, the most probable value of n_i^{ν} , n_i^{*} , are given by

$$\ln \frac{n_i^{*} + g_i}{n_i^{*}} = \alpha + \beta\varepsilon_i$$

or

$$\frac{n_i^{*}}{g_i} = \left[e^{\alpha + \beta\varepsilon_i} - 1\right]^{-1} .$$

Solution to Problem 8-5

a) From Problem 8-1 and 8-2 we have

$$\ell n\ \Omega = \sum_i \left\{ \pm g_i\ \ell n\ \frac{g_i \pm n_i^*}{g_i} + n_i^*\ \ell n\ \frac{g_i \pm n_i^*}{n_i^*} \right\} \qquad \frac{BE}{FD}$$

with

$$\frac{n_i^*}{g_i} = \left[e^{\alpha+\beta\varepsilon_i} \mp 1 \right]^{-1} . \qquad \frac{BE}{FD} \qquad (1)$$

Using the expression

$$S = \underline{\underline{k}}\ \ell n\ \Omega\ ,$$

we obtain

$$S/\underline{\underline{k}} = \sum_i \left\{ \mp g_i\ \ell n\ [1 \mp \exp\ (-\alpha-\beta\varepsilon_i)] + n_i^*\ (\alpha+\beta\varepsilon_i) \right\}$$

$$= \sum_i \left\{ \mp g_i\ \ell n\ [1 \mp \exp\ (-\alpha-\beta\varepsilon_i)] \right\} + \alpha N + \beta E\ . \qquad \frac{BE}{FD}$$

b) If we differentiate the above expression with respect to S, holding V and N constant, we obtain

$$\frac{1}{\underline{\underline{k}}} = \sum_i \frac{g_i}{[\exp\ (\alpha+\beta\varepsilon_i) \mp 1]} \left[-\left(\frac{\partial\alpha}{\partial S}\right)_{N,V} - \left(\frac{\partial\beta}{\partial S}\right)_{N,V} \varepsilon_i \right]$$

$$+ N \left(\frac{\partial\alpha}{\partial S}\right)_{N,V} + \left(\frac{\partial\beta}{\partial S}\right)_{n,V} E + \beta \left(\frac{\partial E}{\partial S}\right)_{N,V} \qquad \frac{BE}{FD}\ .$$

Now using $\left(\frac{\partial E}{\partial S}\right)_{N,V} = T$,

we obtain

Problem 8-5 (continued)

$$\frac{1}{\underline{\underline{k}}} = \left(\frac{\partial \alpha}{\partial S}\right)_{N,V} \left[N - \sum_i \frac{g_i}{[\exp(\alpha+\beta\varepsilon_i) \mp 1]}\right]$$

$$+ \left(\frac{\partial \beta}{\partial S}\right)_{N,V} \left[E - \sum_i \frac{g_i\varepsilon_i}{[\exp(\alpha+\beta\varepsilon_i) \mp 1]}\right] \qquad \frac{BE}{FD}$$

$$+ \beta T$$

$$= \beta T \quad .$$

or

$$\beta = \frac{1}{\underline{\underline{k}}T} \quad .$$

c) If we differentiate the expression for the entropy with respect to N , keeping S and V constant, we obtain

$$0 = \sum_i \frac{g_i}{[\exp(\alpha+\beta\varepsilon_i) \mp 1]} \left[-\left(\frac{\partial \alpha}{\partial N}\right)_{S,V} - \left(\frac{\partial \beta}{\partial N}\right)_{S,V} \varepsilon_i\right]$$

$$+ \left(\frac{\partial \alpha}{\partial N}\right)_{S,V} N + \alpha + \left(\frac{\partial \beta}{\partial N}\right)_{S,V} E + \beta\left(\frac{\partial E}{\partial N}\right)_{S,V} \quad , \qquad \frac{BE}{FD}$$

or

$$0 = \alpha + \beta\left(\frac{\partial E}{\partial N}\right)_{S,V} \quad ,$$

or $\alpha = -\dfrac{\mu}{\underline{\underline{k}}T}$.

Problem 8-5 (continued)

d) If α is very large

$$\frac{n_i^*}{g_i} \approx e^{-\alpha-\beta\varepsilon_i}$$

and $\sum n_i^* = N = \sum g_i e^{-\alpha-\beta\varepsilon_i}$,

or

$$e^{\alpha} = \frac{1}{N} \sum g_i e^{-\beta\varepsilon_i} = \frac{Z}{N} \quad ,$$

or

$$-\frac{\mu}{kT} = \alpha = \ell n \frac{Z}{N} \ .$$

Notice that the quantum-mechanical partition function

$$\ell n\, Z_{Q.M.} = N\alpha \pm \Sigma\, g_i\, \ell n(1 \pm e^{-\alpha-\beta\varepsilon_i})$$

$$\approx N\alpha + e^{-\alpha} \Sigma\, g_i\, e^{-\beta\varepsilon_i}$$

for large values of α .

Using the expressions for α obtained above we may write

$$\ell n\, Z_{Q.M.} = N\alpha + N = N\ell nZ - N\ell nN + N$$

$$= \ell n \left(\frac{Z^N}{N!}\right)$$

Solution to Problem 8-16

Equation (8-52a) is

$$N = \frac{V}{h^3}(2\pi mkT)^{3/2}\left[\frac{1}{z} + \frac{1}{2^{3/2}z^2} + \frac{1}{3^{3/2}z^3} + \dots\right] .$$

If we set $z^{-1} = a_1 n + a_2 n^2 + a_3 n^3 + \dots$

and write

$$n = \left(\frac{2\pi mkT}{h^2}\right)^{3/2}\left[(a_1 n + a_2 n^2 + a_3 n^3 + \dots) + \frac{1}{2^{3/2}}(a_1 n + a_2 n^2 + \dots)^2 + \dots\right]$$

then, upon comparing equal powers of n, we obtain

$$a_1 = \left(\frac{h^2}{2\pi mkT}\right)^{3/2} \equiv \lambda^3$$

$$a_2 = -\frac{1}{2^{3/2}}\lambda^6$$

$$a_3 = \left(\frac{1}{4} - \frac{1}{3^{3/2}}\right)\lambda^9 .$$

We may write the energy, E, as

$$E = \frac{3}{2}kTV\left(\frac{2\pi mkT}{h^2}\right)^{3/2}\left[(a_1 n + a_2 n^2 + \dots) + \frac{1}{2^{5/2}}(a_1 n + a_2 n^2 + \dots)^2 + \dots\right]$$

or, using the values for a_1, a_2, a_3 obtained above, we see that

$$E = \frac{3}{2}kTV\lambda^{-3}\left[\lambda^3 n - \frac{1}{4\sqrt{2}}\lambda^6 n^2 + \left(\frac{1}{8} - \frac{2}{9\sqrt{3}}\right)\lambda^9 n^3 + \dots\right] .$$

Since $PV = \frac{2}{3}E$, we obtain

$$\frac{Pv}{kT} = 1 - \frac{1}{4\sqrt{2}}\frac{\lambda^3}{v} + \left(\frac{1}{8} - \frac{2}{9\sqrt{3}}\right)\left(\frac{\lambda^3}{v}\right)^2 + \dots$$

where

$$v = n^{-1} .$$

Solution to Problem 9.1

We begin by giving the complete solution for the chain with periodic boundary conditions. We assume solutions of the form

$$y_n = A \cos 2\pi(\nu T - fna)$$

where

$$|f| = \frac{1}{\lambda}$$

The periodicity condition requires that

$$y_n = y_{n+N}$$

or

$$A \cos 2\pi(\nu t - fna) = A \cos 2\pi(\nu t - fna - Nfa) .$$

Consequently

Nfa must be an integer

$$Nfa = \ell$$

The possible modes are all included if

$$-N/2 \leq \ell \leq N/2$$

The negative values of ℓ correspond to waves propagating in opposite directions around the chain to those with positive values of ℓ .

The allowed values of the wavelength are given by

$$\lambda_\ell = \frac{Na}{|\ell|}$$

and the allowed frequencies are

$$\nu_\ell = \frac{1}{\pi}\left(\frac{k}{m}\right)^{\frac{1}{2}} \sin \frac{\pi|\ell|}{N}$$

The short-wavelength modes correspond to large values of $|\ell|$ and we may now consider a few of them.

There is one mode with frequency

$$\nu_{max} = \frac{1}{\pi}\left(\frac{k}{m}\right)^{\frac{1}{2}}$$

Problem 9.1 (continued)

corresponding to $\ell = \pm N/2$.

There are two modes with frequency

$$\nu = \frac{1}{\pi}\left(\frac{k}{m}\right)^{\frac{1}{2}} \sin\frac{\pi}{2}\left(1 - \frac{2}{N}\right)$$

corresponding to $|\ell| = \frac{N}{2} - 1$ and so on.

Consider now the chain with fixed boundaries. The suggested solution leads to

$$y_1 = A \sin 2\pi\nu t + B \sin 2\pi\nu t = 0 ,$$

or

$$A = -B ,$$

and

$$y_{N+1} = A\left[\sin 2\pi\left(\nu t - \frac{Na}{\lambda}\right) - \sin 2\pi\left(\nu t + \frac{Na}{\lambda}\right)\right] = -2A \cos 2\pi\nu t \sin\frac{2\pi Na}{\lambda} .$$

The wavelengths of the vibrations are now determined by the condition that

$$\frac{2Na}{\lambda} = j$$

where j is an integer

$$0 \lesssim j \lesssim N .$$

The corresponding frequencies are

$$\nu_j = \frac{1}{\pi}\left(\frac{k}{m}\right)^{\frac{1}{2}} \sin\frac{\pi j}{2N} .$$

The short-wavelength modes correspond to large values of j . Here we list the first few

$$\nu_N = \nu_{max} = \frac{1}{\pi}\left(\frac{k}{m}\right)^{\frac{1}{2}} \qquad j = N$$

$$\nu_{N-1} = \frac{1}{\pi}\left(\frac{k}{m}\right) \sin\frac{\pi}{2}\left(1 - \frac{1}{N}\right) .$$

Notice that $\nu_{N-1} = \nu_{max} + O(1/N^2)$,

$$\nu_{N-2} = \frac{1}{\pi}\left(\frac{k}{m}\right)^{\frac{1}{2}} \sin\frac{\pi}{2}\left(1 - \frac{2}{N}\right)$$

and so on.

Problem 9.1 (continued)

One can see that for these modes the difference between the fixed and periodic chain is that two very closely spaced modes in the fixed chain correspond to two equal modes in the periodic chain. This suggests that for large chains the distribution of frequencies ought to be the same for both chains. To investigate this further we consider modes with wavelengths larger than the crystal spacing $(\lambda >> a)$ but much smaller than the size of the chain $(\lambda << Na)$.

For the periodic chain this leads to

$$\nu_\ell \approx \left(\frac{k}{m}\right)^{\frac{1}{2}} \frac{|\ell|}{N} \qquad 1 < |\ell| < \frac{N}{2}$$

and for the fixed chain, we have

$$\nu_j = \left(\frac{k}{m}\right)^{\frac{1}{2}} \frac{j}{2N} \qquad 1 < j < N \; .$$

The number of modes of the periodic chain with frequency in the interval ν to $\nu + d\nu$ is

$$g(\nu)d\nu = 2d|\ell| = 2N\left(\frac{m}{k}\right)^{\frac{1}{2}} d\nu \; ,$$

the additional factor of 2 arising from the fact that for every $|\ell|$ there are two modes, each propagating in different directions around the chain.

For the chain with fixed boundaries

$$g(\nu)d\nu = dj = 2N\left(\frac{m}{k}\right)^{\frac{1}{2}} d\nu \; .$$

Here we see that the frequency distribution is the same for the periodic and fixed chains for wavelengths which are small compared to the size of the chain and large compared to the lattice spacing.

Solution to Problem 9-6

From thermodynamic relations we may express TdS in two ways, viz:

$$TdS = c_V\, dT + T\left(\frac{\partial P}{\partial T}\right)_V dV$$

$$TdS = c_P\, dT - T\left(\frac{\partial V}{\partial T}\right)_P dP\ .$$

Upon equating these expressions, we obtain

$$dT = \frac{T\left(\frac{\partial P}{\partial T}\right)_V}{c_P - c_V}\, dV + \frac{T\left(\frac{\partial V}{\partial T}\right)_P}{c_P - c_V}\, dP\ .$$

Hence

$$\left(\frac{\partial T}{\partial V}\right)_P = T\left(\frac{\partial P}{\partial T}\right)_V (c_P - c_V)^{-1}\ ,$$

or

$$(c_p - c_v) = T\left(\frac{\partial P}{\partial T}\right)_V \left(\frac{\partial V}{\partial T}\right)_P\ ,$$

but

$$\left(\frac{\partial P}{\partial T}\right)_V = -\left(\frac{\partial P}{\partial V}\right)_T \left(\frac{\partial V}{\partial T}\right)_P\ ,$$

therefore

$$c_P - c_V = -T\left(\frac{\partial V}{\partial T}\right)_P^2 \left(\frac{\partial P}{\partial V}\right)_T$$

$$= \frac{TV\alpha^2}{\kappa}\ .$$

The difference in the specific heats per mole is

$$c_P - c_V = \frac{v_m \alpha^2 T}{\kappa}$$

Solution to Problem 9-8

In analogy with the development in Section 9.32, we may express the frequencies for a two dimensional crystal as

$$\nu = \frac{c}{L}\left(n_1^2 + n_2^2\right)^{1/2} = \frac{c}{L} n \quad ,$$

and the distribution of frequencies $g(\nu)$ is determined by

$$g(\nu)\, d\nu = 2\pi L^2\left(\frac{1}{c_t^2} + \frac{1}{c_\ell^2}\right) \nu d\nu = \frac{4\pi L^2}{c^2} \nu d\nu \quad .$$

The maximum frequency ν_m is given by the relation

$$2N = \int_0^{\nu_m} g(\nu)\, d\nu = 2\pi\left(\frac{L}{c}\right)^2 \nu_m^2 \quad ,$$

or

$$\nu_m = \left(\frac{Nc^2}{\pi L^2}\right)^{1/2} .$$

The energy of the two dimensional solid is

$$E = -\frac{\partial}{\partial\beta}\int_0^{\nu_m} [\ell n\ Z\ (\nu)]\ g(\nu)\ d\nu$$

$$= -\frac{\partial}{\partial\beta}\int_0^{\nu_m}\left[\frac{1}{2}\beta \underline{\underline{h}}\nu - \ell n\ (1 - \ell^{-\beta\underline{\underline{h}}\nu})\right] g(\nu)\ d\nu$$

$$= -\ 4\pi\left(\frac{L}{c}\right)^2 \int_0^{\nu_m}\left[\frac{1}{2}\underline{\underline{h}}\nu - \frac{\underline{\underline{h}}\nu}{e^{\beta\underline{\underline{h}}\nu} - 1}\right]\nu\ d\nu \quad .$$

The specific heat C_V is given as

$$C_V = \frac{4\pi}{\underline{\underline{k}}T^2}\left(\frac{L}{c}\right)^2 \int_0^{\nu_m} \frac{\underline{\underline{h}}^2\nu^3 e^{\beta\underline{\underline{h}}\nu}}{(e^{\beta\underline{\underline{h}}\nu} - 1)^2}\ d\nu$$

$$= 4\pi\left(\frac{L}{c}\right)^2 \frac{\underline{\underline{k}}^3T^2}{\underline{\underline{h}}^2}\int_0^{\xi_m} \frac{\xi^3 e^{\xi}}{(e^{\xi}-1)^2}\ d\xi \quad ,$$

Problem 9-8 (continued)

with $\xi_m = \frac{h\nu_m}{kT}$.

For low temperatures $\xi_m \to \infty$ and the integral approaches a constant value. Thus, for low temperatures

$$C_V \simeq 4\pi\left(\frac{L}{c}\right)^2 \frac{k^3T^2}{h^2} \int_0^\infty \frac{\xi^3 e^\xi}{(e^\xi-1)^2}\, d\xi \quad .$$

Solution to Problem 9.12

The specific heat of a Debye solid is given by

$$C_v = \frac{9\underline{\underline{R}}}{\xi_m^3} \int_0^{\xi_m} \xi^4 \, e^{\xi} \, (e^{\xi} - 1)^{-2} \, d\,\zeta \quad .$$

For $T \gg \theta_D$, $\xi_m = \theta_D/T \ll 1$ and we can expand the integrand in powers of ξ . In this way we find

$$C_v = \frac{9\underline{\underline{R}}}{\xi_m^3} \int_0^{\xi_m} (\xi^2 - \frac{1}{12}\,\xi^4 + \ldots)\, d\,\xi$$

$$= 3\,\underline{\underline{R}} \left[1 - \frac{1}{20} \left(\frac{\theta_D}{T} \right)^2 + \ldots \right] \quad .$$

Solution to Problem 10.11

We treat the earth and sun as disc-shaped black-body emitters of diameter D' and D respectively, separated by a distance R. The earth subtends a solid angle A'/R^2 at the sun where

$$A' = \pi D'^2/4 .$$

Under steady-state conditions, the power emitted by the earth through radiation is equal to the power absorbed by the earth from the sun. Therefore, if we use the Stefan-Boltzmann law for black-body radiation, we may write

$$(2A')\sigma T'^4 = 2A(\sigma T^4)\, A'/4\pi R^2$$

where T' is the steady-state temperature of the earth, and where A is the area of the solar disc, $\pi D^2/4$.

Solving for T' we obtain

$$T' = \frac{T}{2}(D / R)^{\frac{1}{2}} .$$

Using the values

$$T = 5800 \text{ K}$$

$$D = 1.4 \times 10^9 \text{m}$$

and

$$R = 1.5 \times 10^{11} \text{m}$$

we obtain

$$T' = 280 \text{ K} .$$

Solution to Problem 10.16

We use the result quoted in Problem 10.14 that the rate at which energy is transferred from one plate to another is

$$q_{ij} = \frac{\sigma(T_i^4 - T_j^4)}{\varepsilon_i^{-1} + \varepsilon_j^{-1} - 1} .$$

Therefore the rate of energy transfer between the plate at T_1 and the shield of temperature T is

$$q_{13} = \frac{\sigma(T_1^4 - T^4)}{1 + \varepsilon^{-1} - 1} = \varepsilon\sigma(T_1^4 - T^4) .$$

Also the rate of energy transfer between the shield and the plate at T_2 is

$$q_{32} = \varepsilon\sigma(T^4 - T_2^4) .$$

Under steady-state conditions

$$q_{13} = q_{32} ,$$

or

$$T_1^4 - T^4 = T^4 - T_2^4 ,$$

or

$$T = [\tfrac{1}{2}(T_1^4 + T_2^4)]^{1/4} .$$

Substituting the values $T_1 = 800$ K and $T_2 = 300$ K , we obtain

$$T = [\tfrac{1}{2}(800^4 + 300^4)]^{1/4} \approx 676 \text{ K} .$$

The net radiation transfer from the surface at T_1 to the surface at T_2 in the presence of the shield is

$$q'_{12} = \varepsilon\sigma(T_1^4 - T^4) ,$$

Problem 10.16 (continued)

since this is the rate at which energy leaves plate 1 and arrives at plate 2.

Hence

$$q'_{12} = \varepsilon\sigma(T_1^4 - \frac{1}{2}(T_1^4 + T_2^4)) = \frac{1}{2}\varepsilon\sigma(T_1^4 - T_2^4) .$$

In the absence of the shield

$$q_{12} = \sigma(T_1^4 - T_2^4) .$$

Therefore

$$\frac{q'_{12}}{q_{12}} = \frac{\varepsilon}{2} .$$

Inserting the appropriate values we find that if

$$\varepsilon = 0.03$$

we have $q'_{12} = 0.015\ q_{12}$,

or a 98.5 % reduction.

With

$$\varepsilon = 0.6 ,$$

we see that

$$\frac{q'_{12}}{q_{12}} = 0.3 ,$$

or a 70 % reduction.

Solution to Problem 10.21

If we neglect the process of induced emission in radiative transitions, we obtain the rate-balance equation which reads

$$B_{12}n_1u(21,T) = A_{21}n_2 ,$$

or

$$u(21,T) = \frac{A_{21}}{B_{12}}\frac{n_2}{n_1} = \frac{A_{21}}{B_{12}} \exp(-h\nu_{21}/kT) .$$

This is Wien's formula, given by equation (10.44). Here we identify the ratio A_{21}/B_{12} with the quantity $8\pi h\nu_{21}^3/c$, as in the more general case.

Solution to Problem 11.8

The magnetization M is given as

$$M = N g_j \mu_B \, j \, B_j(x)$$

with $x = g_j \mu_B \mu_o H/kT$.

If $B = 16\ W/m^2$

$g_j = 2$

$j = 7/2$,

then $$x = \frac{(2)\ (9.27 \times 10^{-24})\ (16)}{(1.38 \times 10^{-23})\ (4)} = 5.37 \approx 5.4 \ .$$

Under these circumstances $B_{7/2}(5.4) \approx 1$

and

$$M = (6.02 \times 10^{26})\ (2)\ (9.27 \times 10^{-24})\ (7/2)\ j/T\ kgmol$$

$$= 3.91 \times 10^{4}\ Am^2/kmol.$$

The magnetic susceptibility χ is

$$\chi = \frac{M}{VH} = \frac{2(3.91 \times 10^{4})\ (4\pi \times 10^{-7})}{(747 \times 10^{3})\ (3.01)^{-1}\ 10^{-6}\ (16)} = 2.48 \times 10^{-2} \ .$$

Here χ is dimensionless. The factor of two in the expression for χ is due to the fact that one mole of gadolinium sulfate contains two moles of gadolinium ions.

Solution to Problem 11.14

The number of ions dN whose magnetic moments make an angle θ (in the range $d\theta$) with the direction of the magnetic field is given by classical mechanics as

$$dN = \frac{N \sin\theta \exp(\beta\mu_o\mu H \cos\theta)\, d\theta}{\int_o^\pi \sin\theta \exp(\beta\mu_o\mu H \cos\theta)\, d\theta} .$$

The magnetization is

$$M = N\mu \frac{\int_o^\pi \cos\theta \sin\theta \exp(\beta\mu_o\mu H \cos\theta)\, d\theta}{\int_o^\pi \sin\theta \exp(\beta\mu_o\mu H \cos\theta)\, d\theta}$$

$$= N\mu \int_{-1}^{1} ue^{tu}du \Big/ \int_{-1}^{1} e^{tu}du ,$$

with

$$t = \beta\mu\mu_o H ,$$

or

$$M = N\mu \left[\coth t - \frac{1}{t}\right] = N\mu L\left(\frac{\mu_o\mu H}{kT}\right) ,$$

where $L(t) = \coth t - t^{-1}$ is the Langevin function. To obtain this limit from the quantum formula, we have to set $h \to 0$ but to require that the angular momentum $jh/2\pi$ remain finite. This can only be accomplished if we simultaneously require that $j \to \infty$ in such a way that jh remains finite.

Solution to Problem 12.4

a) The number of molecules that impinge on a unit area of the walls of the vessel in one second is $a = nu/4$. Using the values quoted in part (a) we obtain for the number of particles impinging on an area of $10^{-4}\ m^2$

$$aA = (2 \times 10^{25}) \left[\frac{(1.38 \times 10^{-23})\ (273)}{2\pi\ (3.34 \times 10^{-27})}\right]^{\frac{1}{2}} 10^{-4}$$

$$= 8.5 \times 10^{23} \text{ particles per second.}$$

b) Let a_o be the number of molecules striking a unit area of the walls per unit time with a velocity greater than C_o. Then a_o is given by

$$a_o = n\left(\frac{m}{2\pi kT}\right)^{3/2} \int_{C_o}^{\infty} \int_{o}^{\pi/2} \int_{o}^{2\pi} v^3 \sin\theta \cos\theta\, e^{-mv^2/2kT}\, dv\, d\theta\, d\phi$$

$$= n\pi\left(\frac{m}{2\pi kT}\right)^{3/2} \int_{C_o}^{\infty} v^3\, e^{-\frac{mv^2}{2kT}}\, dv\ ,$$

or

$$a_o = n\left(\frac{kT}{2\pi m}\right)^{\frac{1}{2}} \left[1 + \frac{mC_o^2}{2kT}\right] e^{-mC_o^2/2kT}\ .$$

If $C_o = 12{,}000$ m/sec

$$\frac{mC_o^2}{2kT} = \frac{(3.34 \times 10^{-27})\ (12{,}000)^2}{(2)\ (1.38 \times 10^{-23})\ (273)} \approx 64\ ,$$

then

$$a_o = a(65)\ \exp(-64)\ .$$

Using $\exp(-64) \doteq 1.6 \times 10^{-28}$

Problem 12.4 (continued)

we obtain

$$a_o = 8.5 \times 10^{27} \times 65 \times 1.6 \times 10^{-28} = 88 \text{ particles/sec m}^2 .$$

Solution to Problem 12.12

a) If we ignore the dimensions of the sphere and ignore any effects of gravity of the sphere, we may write the collision frequency of the sphere with the piston as

$$\nu = c/2\ell \text{ per second.}$$

If the collisions with the piston are perfectly elastic, then on each collision a momentum 2mc is transferred to the piston. Hence the rate of momentum transfer, or the average force on the piston is $(2mc)\, c/2\ell = mc^2/\ell$. The pressure on the piston is $mc^2/\ell A = \frac{mc^2}{V}$ where A is the cross-sectional area of the piston and V is the enclosed volume. In other words $PV = mc^2$ and this is analogous to the perfect-gas equation of state.

The condition for equilibrium of the piston is

$$Mg = \frac{mc^2}{\ell}$$

or $\ell = mc^2/Mg$.

b) If we now take into account the diameter of the sphere, the collision frequency is modified to

$$\nu = c/2(\ell-2r) .$$

The pressure is now

$$P = \frac{(2mc)c}{2(\ell-2r)A}$$

or $P(V-V_o) = mc^2$

with $V_o = 2rA$.

This is analogous to the van der Waals equation with the excluded-volume corrections.

Problem 12.12 (continued)

c) This piston is slowly withdrawn with a velocity $v << c$. We can show that at each rebounding from the piston, the sphere loses a kinetic energy

$$\Delta E \approx -(2mc)\,v$$

To show this we employ the conservation of momentum and energy relations as

$$mc + Mv = mc' + Mv'$$

and $$\frac{1}{2}m(c'^2 - c^2) = \frac{1}{2}M(v^2 - v'^2) .$$

Combining these two equations we obtain

$$c' + c = v + v' \approx 2v . \qquad \text{Since } v' \approx v .$$

Consequently

$$\Delta E = \frac{1}{2}m(c'^2 - c^2) \approx \frac{1}{2}m((2v-c)^2 - c^2)$$

$$\approx -2mcv ,$$

or

$$\Delta E = -2mcv = \frac{-2mc}{A}\left(\frac{\Delta x}{\Delta t}\right)A = -P\Delta V .$$

Here $P = \dfrac{+2mc}{A\Delta t}$

is the momentum transferred to the piston per unit area per unit time.

Solution to Problem 12.22

The molecular diameter, σ , has been estimated in Problem 2.19. The result obtained there (see p. 19 of these notes) is

$$\sigma \approx 3.3 \times 10^{-8} \text{ cm} .$$

The mass of one molecule is

$$m = 40 \left(6.02 \times 10^{23}\right)^{-1} = 6.65 \times 10^{-23} \text{ gm}$$

and

$$\underline{\underline{k}}T = (293)\ (1.38 \times 10^{-16}) = 4.04 \times 10^{-14} \text{ erg} .$$

Using equation (12.41b) we obtain

$$\eta = \frac{1}{2}\, n u \ell m = \left(\frac{m\underline{\underline{k}}T}{\pi}\right)^{1/2} (\pi\sigma^2)^{-1}$$

$$\approx 27.0 \times 10^{-5} \text{ dyne sec/cm}^2 .$$

Solution to Problem 14-4

Equation (14.18) may be applied to this problem. We shall use T and V as independent variables and set $\Delta N = 0$. Then we write

$$\Delta S = \left(\frac{\partial S}{\partial T}\right)_V^* \Delta T + \left(\frac{\partial S}{\partial V}\right)_T^* \Delta V$$

and

$$\Delta P = \left(\frac{\partial P}{\partial T}\right)_V^* \Delta T + \left(\frac{\partial P}{\partial V}\right)_T^* \Delta V \ .$$

Consequently

$$\Delta P \Delta V - \Delta S \Delta T = \left(\frac{\partial P}{\partial V}\right)_T^* (\Delta V)^2 - \left(\frac{\partial S}{\partial T}\right)_V^* (\Delta T)^2$$

$$+ (\Delta T \Delta V) \left[\left(\frac{\partial P}{\partial T}\right)_V^* - \left(\frac{\partial S}{\partial V}\right)_T^*\right]$$

but $\left(\frac{\partial P}{\partial T}\right)_V^* = \left(\frac{\partial S}{\partial V}\right)_T^*$.

Therefore the probability of a simultaneous fluctuation in V and T is

$$P = C \exp \frac{1}{2\underline{\underline{k}}T} \left[\left(\frac{\partial P}{\partial V}\right)_T^* (\Delta V)^2 - \left(\frac{\partial S}{\partial T}\right)_V^* (\Delta T)^2\right] \ .$$

To obtain C we integrate over all values of ΔV and ΔT . We therefore obtain

$$P = \frac{1}{2\pi \underline{\underline{k}}T} \left[\frac{1}{V\kappa_T} \left(\frac{\partial S}{\partial T}\right)_V^*\right]^{\frac{1}{2}} \exp - \frac{1}{2\underline{\underline{k}}T} \left[\frac{(\Delta V)^2}{V\kappa_T} + \left(\frac{\partial S}{\partial T}\right)_V^* (\Delta T)^2\right]$$

where $\kappa_T = -\frac{1}{V}\left(\frac{\partial V}{\partial P}\right)_T^*$ is the isothermal compressibility.

We see immediately that fluctuations in V and T are statistically independent.

Problem 14-4 (continued)

To compute $(\Delta E)^2$ we use

$$\Delta E = \left(\frac{\partial E}{\partial V}\right)_T^* \Delta V + \left(\frac{\partial E}{\partial T}\right)_V^* \Delta T$$

and

$$\left\langle(\Delta E)^2\right\rangle = \left(\frac{\partial E}{\partial V}\right)_T^{*2} \left\langle(\Delta V)^2\right\rangle + c_V^2 \left\langle(\Delta T)^2\right\rangle .$$

The quantities $\left\langle(\Delta V)^2\right\rangle$ and $\left\langle(\Delta T)^2\right\rangle$ may be obtained as

$$\left\langle(\Delta V)^2\right\rangle = \frac{1}{2\pi \underline{\underline{k}}T}\left[\frac{1}{V\kappa_T}\left(\frac{\partial S}{\partial T}\right)_V^*\right]^{\frac{1}{2}} \int_{-\infty}^{\infty} (\Delta V)^2 \exp \frac{-(\Delta V)^2}{2\underline{\underline{k}}TV\kappa_T}\, d(\Delta V)$$

$$\int_{-\infty}^{\infty} \exp\left[-\frac{1}{2\underline{\underline{k}}T}\left(\frac{\partial S}{\partial T}\right)_V^* (\Delta T)^2\right] d(\Delta T)$$

$$= \underline{\underline{k}}TV\kappa_T = -\underline{\underline{k}}T\left(\frac{\partial V}{\partial P}\right)_T^* .$$

Similarly

$$\left\langle(\Delta T)^2\right\rangle = \frac{\underline{\underline{k}}T^2}{c_V} .$$

Consequently

$$\left\langle(\Delta E)^2\right\rangle = \underline{\underline{k}}T\left[Tc_V - \left(\frac{\partial E}{\partial V}\right)_T^{*2}\left(\frac{\partial V}{\partial P}\right)_T^*\right] .$$

We also note that

$$\left\langle\Delta E \Delta V\right\rangle = \left(\frac{\partial E}{\partial V}\right)_T^* \left\langle(\Delta V)^2\right\rangle$$

and that

$$\left(\frac{\partial E}{\partial V}\right)_T^* = T\left(\frac{\partial P}{\partial T}\right)_V^* - P .$$

Problem 14-4 (continued)

Therefore

$$\langle(\Delta E \Delta V)\rangle = \underline{\underline{k}}T \left[T\left(\frac{\partial V}{\partial T}\right)_P^* + P\left(\frac{\partial V}{\partial P}\right)_T^* \right] ,$$

where we have used

$$\left(\frac{\partial P}{\partial T}\right)_V^* \left(\frac{\partial V}{\partial P}\right)_T^* = -\left(\frac{\partial V}{\partial T}\right)_P^* .$$

For a monatomic ideal gas

$$PV = N\underline{\underline{k}}T \quad ; \quad E = \frac{3}{2} N\underline{\underline{k}}T$$

$$c_V = \frac{3}{2} N\underline{\underline{k}} .$$

Here

$$\langle(\Delta E)^2\rangle = \underline{\underline{k}}T\left(\frac{3}{2}N\underline{\underline{k}}T\right) = \frac{2}{3N} E^2$$

$$\left\langle\frac{(\Delta E)^2}{E^2}\right\rangle^{\frac{1}{2}} \sim \frac{1}{\sqrt{N}} .$$

Also $\langle(\Delta V)^2\rangle = V^2/N$

and $\langle \Delta E \Delta V \rangle = 0$.

Solution to Problem 14.10

In this example we may use equation (14.18) with $\Delta V = \Delta T = 0$,

$$\Delta\mu = \left(\frac{\partial\mu}{\partial N}\right)^*_{T,V} \Delta N ,$$

and $\left(\frac{\partial\mu}{\partial N}\right)^*_{T,V} = \frac{V}{N^2\kappa_T}$

with $\kappa_T = \frac{-1}{V}\left(\frac{\partial V}{\partial P}\right)^*_{T,N}$.

Hence

$$P = C \exp\left[-\frac{1}{2\underline{\underline{k}}T}\frac{V}{N^2\kappa_T}(\Delta N)^2\right] .$$

After normalizing P we find that

$$C = \left(\frac{V}{2\underline{\underline{k}}TN^2\pi\kappa_T}\right)^{\frac{1}{2}} ,$$

and that

$$\left\langle(\Delta N)^2\right\rangle = \underline{\underline{k}}T\,\frac{N^2}{V}\,\kappa_T .$$

Consequently the relative fluctuations in the number of particles in the system are

$$\left[\left\langle\left(\frac{N - \langle N\rangle}{\langle N\rangle}\right)^2\right\rangle\right]^{\frac{1}{2}} = \left[\frac{\underline{\underline{k}}T\kappa_T}{V}\right]^{\frac{1}{2}} .$$

For a perfect gas

$$\kappa_T = \frac{-1}{V}\left(\frac{\partial V}{\partial P}\right)_T = \frac{1}{P}$$

and

Problem 14.10 (continued)

$$\left[\left\langle\left(\frac{N - \langle N\rangle}{\langle N\rangle}\right)^2\right\rangle\right]^{\frac{1}{2}} = \left[\frac{kT}{PV}\right]^{\frac{1}{2}} .$$

For $T = 300$ K ; $P = 1$ atm $= 1.013 \times 10^6$ dyne/cm^2 this quantity takes on the following values.

i) For $V = 1$ cm^3

$$\left[\left\langle\left(\frac{N - \langle N\rangle}{\langle N\rangle}\right)^2\right\rangle\right]^{\frac{1}{2}} \approx 2 \times 10^{-10} .$$

ii) For $V = 10^{-3}$ cm^3

$$\left[\left\langle\left(\frac{N - \langle N\rangle}{\langle N\rangle}\right)^2\right\rangle\right]^{\frac{1}{2}} \approx 2 \times 10^{-10} (10^3)^{\frac{1}{2}} \approx 6 \times 10^{-9} .$$

iii) For $V = 10^{-15}$ cm^3

$$\left[\left\langle\left(\frac{N - \langle N\rangle}{\langle N\rangle}\right)^2\right\rangle\right]^{\frac{1}{2}} \sim (2 \times 10^{-10}) (10^{15})^{\frac{1}{2}} \approx 6 \times 10^{-3}$$

or $\sim 0.6\%$ fluctuation.

iv) For $V = 10^{-21}$ cm

$$\left[\left\langle\left(\frac{N - \langle N\rangle}{\langle N\rangle}\right)^2\right\rangle\right]^{\frac{1}{2}} \approx (2 \times 10^{-10}) (10^{21})^{\frac{1}{2}} = 6.32 .$$

Here the fluctuations are very large and our approximations break down. These fluctuations produce a scattering of light with wavelength of the order of 10^{-7} cm.